Environmental Aspects, Applications and Implications of Nanomaterials and Nanotechnology 2017

AA002845

Topical Conference at the 2017 AIChE Annual Meeting

Minneapolis, Minnesota, USA
29 October - 3 November 2017

ISBN: 978-1-5108-5780-3

Printed from e-media with permission by:

Curran Associates, Inc.
57 Morehouse Lane
Red Hook, NY 12571

Some format issues inherent in the e-media version may also appear in this print version.

Copyright© (2017) by AIChE
All rights reserved.

Printed by Curran Associates, Inc. (2018)

For permission requests, please contact AIChE
at the address below.

AIChE
120 Wall Street, FL 23
New York, NY 10005-4020

Phone: (800) 242-4363
Fax: (203) 775-5177

www.aiche.org

Additional copies of this publication are available from:

Curran Associates, Inc.
57 Morehouse Lane
Red Hook, NY 12571 USA
Phone: 845-758-0400
Fax: 845-758-2633
Email: curran@proceedings.com
Web: www.proceedings.com

TABLE OF CONTENTS

Some abstracts in this book refer to images that were not made available by the publisher

485452 Soot Source Identification By Laser Derivitization

Abstract:

Combustion produced soot is highly variable with details as dependent upon combustion conditions. Our prior studies have shown soot nanostructure to be dependent upon the source via quantification of HRTEM images for nanostructural parameters. In principle this permits identification of the soot source and its contribution to any particular receptor site. Yet many structural aspects are subtle and the chemistry of and between the lamella is unaddressed. How best to bring out small differences in nanostructure and other seemingly subtle differences in chemistry? We proposed the process of pulsed laser annealing to highlight compositional and structural differences thereby distinctively and uniquely identifying the source of the soot. Our overall objective is then to develop the laser-based heating as an analytical tool and identify the process conditions and operational parameters for optimal derivitization. Specific tasks directed towards achieving this goal include 1) Identifying optimal laser operational parameters for derivitization, 2) Defining the dependence upon nanostructure and composition using model soots while also identifying variability and range of products for uniform and homogeneous starting material, 3) Demonstrating differentiation upon combustion derived soots from real engines such as diesel, gasoline, boilers, etc., and 4) Quantifying the nanostructural changes and statistical differences between the derivatives of these actual samples to formulate statistical confidence intervals for nanostructure parameters of the derivatized soots. Results will be presented accordingly.

Copyright © American Institute of Chemical Engineers. All rights reserved.

489838 Evaluation of Biological Effects Using a Nano-Ceria Based Diesel Fuel Additive with *in vitro* Air-Liquid Interface Cell Exposure Systems of Different Flow Patterns

Abstract:

Diesel exhaust is a large contributor of particulate matter in air pollution, which can lead to pulmonary conditions such as chronic obstructive pulmonary disorder and even lung cancer. To reduce emissions, the introduction of metal oxide fuel additives into the engine provides additional oxidation points for soot particles and greenhouse gases. These additives have been shown to increase the number of nanoparticles in the exhaust which when inhaled can deposit in the lower lungs with an extended residence time, allowing more time for particle-cell interactions. Aerosols can be tested for toxicity *in vitro* at the air-liquid interface (ALI), where the cells are grown on a porous membrane, allowing one side to be exposed after equilibration with air, while the other side receives nutrients from liquid media. The biological effects of diesel exhaust have not been extensively studied *in vitro* at the ALI, nor with metal oxide additives.

The objective of this work is to determine the potential cytotoxicity and oxidative activity changes in lung cells cultured at the ALI after exposure to exhaust generated by a commercial, nano-cerium oxide based diesel fuel additive as directed with low-sulfur fuel. Two ALI cell exposure systems with two different flow patterns were utilized. Exhaust was generated by a partly loaded, single cylinder, four-stroke, air cooled, direct inject diesel power generator. The particle concentration was maintained at approximately 10^6 particles/cm^3 as measured by a Scanning Mobility Particle Sizer. Deposition was measured gravimetrically on glass-fiber filters and blank cell culture inserts, qualitatively using Transmission Electron Microscopy, and online during exposure using quartz crystal microbalance measurements. These can be correlated to the biological effects of cytotoxicity and the oxidative activity, including reactive oxidative species generation and oxidative protection.

Acellular experiments suggest a correlation between the deposition and the type of fuel used for the newly designed Portable In Vitro Exposure Cassette. Using diesel exhaust, approximately one half of the filter mass is collected on the insert; whereas for additive fuel, approximately one sixth of the filter mass is collected. Preliminary cellular results suggest a decrease in cytotoxicity and no statistically significant effect on reactive oxygen species generation with the use of the nano-cerium oxide additive.

Copyright © American Institute of Chemical Engineers. All rights reserved.

499118 Oxidative Stress Induced By Ambient Air PM_x: Which Are the Main Sources?

Abstract:

Particulate Matter (PM) can generate reactive oxygen species (ROS) through two different mechanisms. Through the oxidative constituents absorbed onto their surface and can cause oxidation and through their ability to induce cellular generation of ROS in target cells, such as pulmonary epithelial cells and pulmonary macrophages. Chemical components of ambient PM, such as quinones and transition metals can initiate the catalysis of redox reactions in a biological system. Few studies so far have focused on the link between ROS generation with the respective aerosol sources. Identifying sources of redox-active PM_x is important for environmental health and industrial technology development. A variety of methods are available for measuring the capacity of PM_x to catalyse redox reactions. The dithiothreitol (DTT) assay is a commonly used acellular assay due to its ability to correlate well with biological markers. Several source apportionment techniques (e.g., positive matrix factorization and chemical mass balance) have been applied for the identification of the major emission sources associated with PM-catalyzed ROS generation.

The present study aimed at investigating the most important sources that drive the redox activity of ambient PM_x (PM_1, $PM_{2.5}$, PM_{10}) at two sites in Thessaloniki, the 2nd largest city in Greece, capturing different degrees of non-traffic and traffic-related activities. For this reason, an extensive campaign was carried out at two types of locations in the area of Thessaloniki to determine the chemical composition of urban aerosols and to correlate their toxicity with PM sources. From the chemical analysis of the PM_x sampled we found that oxidative potential of PM was not correlated to its mass concentration. On the contrary, a strong correlation between chemical composition and oxidation capacity of PM was observed. A rough estimation of ROS sources was obtained based on correlations between DTT activity and chemical species found in the respective PM samples and on connecting those species with sources. At the traffic site Cu and Zn were correlated with DTT activity, highlighting the presence of vehicular emissions while at the urban background site K, levoglucosan and PAHs were correlated with DTT activity of PM.

Ambient air data showed high levels of PM during the sampling period. However, the contribution of traffic to PM levels in urban air is reduced in the last five years in Greece. Also it was observed that the mean concentrations of PM_x at the urban background site were higher than at the traffic site. Similarly, the average total PAH concentrations were significantly higher at the urban background site. Black carbon (BC) levels (associated to internal combustion sources) did not differ significantly between the two sites. During the sampling period the contribution of biomass-fueled domestic heating to ambient air PM mass concentration was calculated to rise up to 34%. The values of levoglucosan concentrations shed light on the relative contribution of the different sources at the two sampling sites. At the background site, the estimated mean concentrations of PM_1, $PM_{2.5}$ and PM_{10} were 0.7 μg/m^3, 1.1 μg/m^3 and 1.4 μg/m^3 respectively, whereas at the traffic site the corresponding values were 0.3 μg/m^3, 0.4 μg/m^3 and 0.5

$\mu g/m^3$. Regarding the oxidative potential of PM_x, the one corresponding to PM_1 was the highest, while the values for $PM_{2.5}$ and PM_{10} followed. On a mass basis, ultrafine particles were more potent than either the fine or coarse particles. UFPs are highly chemically reactive due to their small size and large surface area and thus more harmful to human health. During the current study, the average mass-normalized redox activity of PM_x measured in PM_1, $PM_{2.5}$ and PM_{10} were 169 ± 90, 118 ± 50, 90 ± 50 pmol min^{-1} μg^{-1}, respectively at the background site while at the traffic site were 135 ± 80 pmol min^{-1} μg^{-1} for PM_1, 100 ± 75 pmol min^{-1} μg^{-1} for $PM_{2.5}$ and 89 ± 50 for PM_{10}.

To identify the contribution of various sources to the oxidative properties of PM_1, $PM_{2.5}$ and PM_{10}, a PMF (positive matrix factorization) analysis was conducted using the DTT activity and measured chemical composition data at each sampling site. The pooled data of PM_x at each sampling site were used to increase the statistical power of PMF. A good stability was shown, for urban background data, by choosing 4 factors while for traffic data 5 factors were selected. At background site 4 factors were determined by PMF. The first had high loadings of crustal elements (Al, Si, Ca, Ti, Mn, Fe) and it was attributed to soil resuspension. The second contained high loadings of BC, Ca, V, Ni, Cu, Zn and Pb and represented vehicular traffic. Transition metals such as Cu, Fe, Mn, and Zn, which have been related to adverse health outcomes, were associated to this factor by 91%, 12 %, 52 % and 64 %, respectively. Mn was also associated with 45% to the soil resuspension factor. Zn was found in a mixture of factors, associated with 11 % with soil resuspension and 22 % with biomass burning. As third factor, secondary aerosols were identified, which were characterized with high loadings of S, BC NO_3^-, benz[a]anthracene and chrysene. BC was found to be associated with 23%, S with 73% and benz[a]anthracene and chrysene 92 and 99 % to secondary aerosols, respectively. The fourth factor had a clear signature of biomass burning with very high concentrations of BC, K and PAHs (BbF, BkF, BeP, BaP, IcdP, BghiP). Biomass burning appears to be a significant contributor to PAHs; the latter are responsible for a large part of its toxic effects. Based on these results, PAHs are important to the oxidative potential of PM due to their ability to be transformed into quinones through photochemical reactions with atmospheric oxidants. At the urban background site biomass burning dominated the ROS-generation potential contributing by 72% to the DTT activity of PM_1, while relatively low contributions were observed from vehicle emissions (25%) and secondary formation (1%). Soil dust made a close to null contribution to DTT activity (2%). The major drivers of the DTT activity of $PM_{2.5}$ at the urban background site were biomass burning and vehicle emissions, with 62% and 34% of contributions, respectively. This was followed by soil dust, which contributed 3%, while secondary aerosol formation contributed minimally (1%) to the DTT activity of $PM_{2.5}$. Moreover, biomass burning and vehicle emissions were the dominant sources of DTT activity of PM_{10} at the urban background site with contributions of 47 and 38 %, respectively, while secondary oxidation processes and soil dust contributed 1% and 14 % to the DTT activity of PM_{10}.

At the traffic site five factors were identified as important by PMF. The first factor contained high loadings of BC, Ca, V, Ni, Cu, Zn and Pb and it was attributed to vehicular emissions. It is noted that V and Ni had high loading in this factor indicating that this factor was mixed with oil combustion. Ni and V originate from oil combustion which is linked to diesel vehicles and shipping emissions. The second factor was biomass burning characterized by very high concentrations of BC, K and PAHs. Secondary aerosol was third with very high concentrations of S, BC. The fourth factor had high loading of crustal elements (Al, Si, Ca, Ti, Mn, Fe) while the fifth factor had high loadings of Mn, Cu, Zn, Ca and Fe. Vehicle emissions were the dominant contributor to DTT activity of PM_x contributing 62% to PM_1, 66 % to $PM_{2.5}$ and 51% to PM_{10}. Biomass burning contributed 15% to the DTT activity of PM_1, 12% to $PM_{2.5}$ and 13% to PM_{10}. Secondary aerosol dominated the ROS-generation potential contributing 17% to DTT activity of PM_1, 16% to DTT activity of $PM_{2.5}$, 13% to DTT activity of PM_{10}. Soil dust had 5% contribution to DTT activity of PM_1, 5% to DTT activity of $PM_{2.5}$, 12% to DTT activity of PM_{10} while road dust contributed 1% to DTT activity of both PM_1 and $PM_{2.5}$ and 11% to DTT activity of PM_{10}.

Vehicle emissions and biomass burning were the two most important sources that drive DTT redox activity of ambient PM_x in Thessaloniki, Greece. Biomass burning was the largest contributor at the urban background site while vehicle emissions at the traffic site.

Copyright © American Institute of Chemical Engineers. All rights reserved.

485453 Development of Predictive Tools for Aviation Non-Volatile Particulate Emissions

Abstract:

Aviation black carbon (BC) emissions impact climate and health. Inventory estimates are essential to quantify these effects. These in turn require a means of estimating BC emission indices from jet aircraft. The first order approximation (FOA3) currently employed to estimate BC mass emissions grossly under predicts BC emissions due to inaccuracies in measuring low smoke numbers (SNs) produced by modern high bypass ratio engines. The recently developed formation-oxidation (FOX) method removes the need for and hence uncertainty associated with (SNs), instead relying upon engine conditions in order to predict BC mass. Using the true engine operating conditions an improved FOX (ImFOX) predictive relation is developed. Still, the current methods are not optimized to estimate cruise emissions or account for the use of alternative jet fuels with reduced aromatic content. Here improved correlations are developed to predict engine conditions and BC mass emissions at ground and cruise altitude. This new ImFOX is paired with the recently developed Approximation for Soot from Alternative Fuels (ASAF) and a newly developed C/H relation to predict emissions from alternative fuels and fuel blends.

Copyright © American Institute of Chemical Engineers. All rights reserved.

502700 Molecular Initiating Events Linked to Carbon Nanotube Functionalization for Medical Applications

Abstract:

The widespread use of functionalized carbon nanotubes (CNTs) makes the understanding of potential harmful effects highly important. Numerous studies have been conducted in the last years on toxicological effects of CNTs. However, existing data are controversial and findings have been difficult to interpret in some cases. In the present study, two cell culture systems, human A549 pneumocytes and HaCaT keratinocytes, were used to assess the modulation of gene expression due to exposure to single and multi-walled CNTs. Moreover, CD-1 male mice were exposed to the CNTs tested by intra-tracheal instillation and lung samples were taken and analyzed after 1 day of exposure. Differentially functionalized CNTs (MW-COOH and MW-NH2) were tested in comparison with pristine multi-wall carbon nanotubes (MWCNTs) and single-wall carbon nanotubes (SWCNTs).

Cells were seeded in 96-well plates at the density of 1x 104/cm^2 in complete medium. After 24 h of cell attachment, plates were washed with 100 l/well of phosphate buffered saline (PBS). Cells were then exposed to suspensions (10 l) of the test materials (pristine MW-CNTs, MW-COOH, MWNH2) at concentration between 1 and 100 Î¼g/ml, for 24 or 48 h. No fetal bovine serum was used in these preparations as it was proven to interact with nanotubes.

Each experiment was repeated independently three times. Per experiment all treatments were repeated six times per 96-well plate. Identical treated cultures were taken as replicate measure for statistical tests. Significant effects ($p < 0.05$) were determined by one-way analyses of variance (ANOVA) (software package SPSS Inc. 1999).

Three groups of four CD-1 male mice were exposed to all tested types of CNTs (both pristine and functionalized) by intra-tracheal installation for one day. SW-CNTs were used at concentrations between 0.1 and 100 Î¼g/ml, while all multiple wall CNTs (both pristine MW-CNTs and functionalized MW-COOH, MW-NH_2) were used at concentrations between 1 and 100 Î¼g/ml. After the exposure period lung samples were taken from the exposed mice and the respective modulation in gene expression was analysed in the whole genome as described below.

Gene expression profiling offers considerable potential for identifying chemical causation of effects induced in exposures to complex mixtures, and for understanding the mechanistic basis for their phenotypic effects. It entails a series of experimental steps, including RNA extraction and preparation, microarray hybridisation, chemiluminescent signal detection, image acquisition and image analysis of the microarrays, and microarray data analysis (in our case we used the Applied Biosystems Expression System software to extract assay signal and signal to noise values from the microarray images). Microarray data analysis identifies differentially expressed genes, applying different statistical filters to allow for only the statistically and biologically significant probes to be evaluated. Significant probe sets were evaluated for relevance to canonical pathway, molecular function and biological function using the Protein ANalysis THrough Evolutionary Relationships (PANTHER) Classification System (http://www.pantherdb.org). Statistically significant over- and under-represented annotation categories were determined by binomial statistics, using the observed number of expressed genes versus the numbers expected by chance within a certain annotation group. Categories with p-values > 10-2 are usually rejected. The results of microarray analysis were validated using quantitative real-time polymerase chain reaction (PCR); the latter determines the expression level of messenger RNA (mRNA) in differentially expressed genes. Transcriptomics analysis of the whole genome in the keratinocytes (HaCaT) selected as *in vitro* models in this study showed that gene expression significantly differs according to the duration of exposure. A large number of genes (between 1200 and 2200) changed their expression level when the cells were exposed for 6 hours to the different types of CNTs analysed, namely SWCNT, MWCNT, MW-NH_2 and MW-COOH. On the contrary, long-lasting exposure (48 hours) resulted in a significant reduction in the number of genes expression differentially compared to the control. Less than 500 genes were expressed differentially in the case of MWCNT, MW-NH2 and MW-COOH; only in the case of SWCNTs ca. 1800 genes showed differential expression after 48 hours of exposure. Similar results were found when we analyzed the data isolating the 25 genes involved in the molecular pathway associated to oxidative stress. The picture changes in terms of modulated gene expression, when focusing on the genes involved in the pathway of inflammation mediated by chemokine and cytokine signalling. SWCNTs and MWCNTs showed a reduction in the number of genes that had modulated their expression level after long-term exposure (48h); on the contrary, both types of functionalized CNTs showed practically no difference in the number of genes with modulated expression when comparing short and long-term exposure data. The observed differences in the level of gene expression as a function of exposure duration may indicate that different cellular mechanisms are involved, namely that short-term response indicates the onset of defensive mechanisms encompassing several molecular pathways and, thus, a large number of genes. The observed effects after 48 h of exposure would seem to indicate the molecular processes that underlie actual cellular toxic response to CNT exposure. The persistence of the induction of genes driving chemokine and cytokine-signalling-mediated inflammation showed that further attention would have to be paid to the inflammatory effects of the tested CNTs, and in particular the functionalized ones. For this reason, we analyzed the gene expression of the genes involved in the inflammation pathway for the two types of in vitro models we used, i.e. HaCaT keratinocytes and A549 lung epithelial cells. Comparative analysis showed that the gene expression of A549 cells was much higher than the gene expression of the keratinocytes after 48 h of exposure.

The *in vitro* observations outlined above indicate that the main pathway to be analyzed in the case of in vivo experiments is inflammation in the lungs mediated by chemokine/cytokine signalling. Identification of specific chemokines involved in inflammation in the lung would result in the discovery of specific biomarkers of effect that may be crucial for the early diagnosis of the toxic potency of CNTs, functionalized or not. The whole genome of lung tissue of CD-1 male mice has been analyzed after 1 day of exposure to CNTs via intra tracheal instillation (1 mg/kg bw).

Similar numbers of genes modulate their expression level after exposure to pristine SW and MW CNTs, whereas significant differences exist between the two functionalized CNTs: MW-NH_2 have a lesser effect than the non-functionalized MWCNTs; on the contrary, MW-COOH affect a much higher number of genes. Another interesting observation is that out of the 180-700 genes with significantly modulated expression only 15 are common across the four treatments.

The microarray results were validated with real-time PCR (polymerase chain reaction). Using bioinformatics analysis on the Integromics® software we made a comparative assessment of the main genes related to chemokine signalling-mediated inflammation. Using a more stingent cut-off criterion for significance of modulation in gene expression (accepting as significant only the genes with more than 3-fold change in expression level) it was clear that MW-COOH had a very significant effect on the expression of all the genes of interest. Using the PANTHER database coupled to the significance analysis using Integromics® we compared the induction of the inflammation mediated by chemokine and cytokine signalling pathway after exposure to MW-COOH vs. MWCNTs and after exposure to MW-NH2 vs. MWCNTs. It was clear that the inflammation pathway was significantly induced by MW-COOH, and it was less so by MW-NH_2 or pristine MWCNTs.

Analysis of the effects of different types of functionalized and pristine CNTs on gene expression showed that both the chemical family and the type of functionalization influence significantly the induction of key toxicity pathways such as inflammation mediated by chemokine and cytokine signalling. In particular, MW-COOH induced the over-expression of chemokine (C-C motif) ligand 2 (Ccl2) and ligand 19 (Ccl19). Both are cytokines involved in immunoregulatory and inflammatory processes. Ccl19 expression was induced significantly also after exposure to MW-NH2 and SWCNTs. Moreover, MW-COOH induced the over-expression of chemokine (C-X-C motif) ligand 10 (Cxcl10) and ligand 11 (Cxcl11), interferon gamma-induced proteins, involved in immunoregulation. More importantly, clinical observations have recently connected the abundant presence of these cytokines to hypersensitivity pneumonitis and, to a lesser extent, to lung fibrosis.

In summary, this study indicated that the biological effect of CNTs at the genomic level could be modified by chemical functionalization in association with changes in solubility and tendency of CNTs to form agglomerates. One of the goals of functionalizing CNTs is to increase their solubility in aqueous media, a feature that may make CNTs more compatible with physiological systems. Differential modulation in gene expression was observed as a function of single or multiple wall geometry and presence of specific functional groups. Comparison of gene expression between in vitro and in vivo exposure to CNTs revealed significant differences in the level of biological response induced towards endpoints such as oxidative stress or inflammation. In particular, MW-COOH induced excessive coding for specific proteins associated clinically with lung fibrosis and other inflammatory processes.

Copyright © American Institute of Chemical Engineers. All rights reserved.

485152 Effect of Combustion Particle Size on Pathologically Important Responses in Lung Cells

Abstract:

Numerous studies have demonstrated associations between combustion-derived particulate matter (cdPM) and adverse human health outcomes. cdPM is a significant contributor to fine and ultrafine PM levels, particularly in urban areas. However, understanding the relationship between cdPM and its health effects is complicated by cdPMâ€™s complex nature: a dynamic mixture of particles and condensed material with different sizes, shapes, and chemical compositions that is generated from a range of sources. Furthermore, several confounding factors make comparing results from various studies difficult, including particle sources and for cdPM: fuel composition, engine age, combustion conditions, lubricants, and sampling methods. For example, experimental and epidemiological studies about the effect of various PM size fractions (PM_{10}, $PM_{2.5}$ and PM_{1}) provide somewhat conflicting results. As particle size diameter decreases, its relative surface area increases, which in turn increases the bioavailability of atoms or molecules leading to, for example, increased pro-inflammatory effects. This study focuses on directly testing the effect of cdPM size on biological responses in lung cells.

The cdPM is generated from a premixed flat-flame burner combusting a jet-fuel surrogate at an equivalence ratio of 2.1. A nitrogen shroud and a glass housing minimize atmospheric effects. The sample is diluted, thermally denuded, and size separated using a differential mobility analyzer. Three different cdPM size fractions are examined (with particle diameters of 25, 45, and 65 nm) for their effect on oxidative potential, cell viability, cellular uptake, inflammatory response, cytochrome P450 (CYP) 1 A1 and 1 B1 mRNA induction, and the activation of transient receptor potential Ankyrin-1 (TRPA1). The biological outcomes are evaluated in A 549 human alveolar basal epithelial cells, with the exception of the TRPA1, which is evaluated in TRP over-expressing HEK293 cells. These results are a step towards understanding some of the interactions between understanding the links between combustion particle aging and health effects.

Copyright © American Institute of Chemical Engineers. All rights reserved.

501134 Oxidative Stress Induced By Secondary Organic Aerosols (SOA) Generated from Biogenic and Anthropogenic Precursors

Abstract:

Secondary organic aerosols (SOA), formed from oxidation of hydrocarbons followed by gas-particle partitioning, constitute a substantial fraction of ambient fine particulate matter (PM). The health implications resulting from exposure to SOA is unclear. In this work, we investigate reactive oxygen and nitrogen species (ROS/RNS) and inflammatory responses to SOA generated from biogenic and anthropogenic precursors. Experiments are performed in the Georgia Tech Environmental Chamber (GTEC) facility, which consisted of two 13m^3 Teflon reactors surrounded by UV and fluorescent sunlights. Hydrocarbons studied include biogenic (isoprene, α-pinene, β-caryophyllene) and anthropogenic (pentadecane, *m*-xylene, and naphthalene) sources. A series of chamber experiments is designed to probe the 1) effect of NO_x (RO_2+HO_2 vs. RO_2+NO reactions), 2) effect of relative humidity, and 3) presence of transition metals, on the ROS/RNS and inflammatory responses of different SOA generated. Murine alveolar macrophages were used to measure intracellular ROS/RNS production. The secreted levels of cytokines such as tumor necrosis factor-α (TNF-α) and interleukin-6 (IL-6) are also measured. We find that SOA precursor identity, reaction condition, and chemical composition affect all measured responses in a hydrocarbon-specific manner. TNF-α levels reach a plateau with increasing IL-6 levels; naphthalene SOA is an exception to this trend and it induces higher levels of TNF-α and ROS/RNS than other SOA. The correlations between SOA chemical composition (bulk elemental ratios) and cellular response will be discussed. Overall, results from this study suggest that the health effects of SOA are important when considering the health impacts of ambient PM.

Copyright © American Institute of Chemical Engineers. All rights reserved.

501657 Quantitative Analysis of the Deposited Nanoparticle Dose on Cell Cultures By Optical Absorption Spectroscopy

Abstract:

Nanoparticles made by aerosol-based technologies find a wide range of promising applications (Pratsinis, 2010), though concerns for their potential health impact exist (Xia et al, 2009). Upon in vitro cytotoxicity evaluations of such particles, agglomeration and settling need to be considered as they influence directly the particle dose that is deposited on the employed cell cultures (Limbach et al, 2005).

Here, nanoparticles of various chemical compositions and sizes are made by flame spray pyrolysis (Madler et al, 2002) and their in vitro cytotoxicity against murine macrophages is evaluated (Sotiriou et al, 2014), accounting for their agglomeration and settling in cell culture media. The agglomerate size distributions of the particles in suspension are measured by dynamic light scattering (Wengeler et al, 2006) and their settling is evaluated by ultraviolet-visible spectroscopy (Sotiriou and Pratsinis, 2010). The cell viability after 24 h incubation with such particle suspensions is monitored by a tetrazolium salt reduction assay in both upright (Sotiriou et al, 2014) and inverted cell culture configurations (Cho et al, 2011, Lee et al, 2014) (Figure 1).

Figure 1. Relative viability of RAW 264.7 murine macrophages after 24 h incubation with nanosilver or nanogold suspensions as a function of nominal metal mass concentration in standard upright and inverted cell culture configurations.

For given nominal particle mass concentrations, three categories of nanoparticles can be distinguished, based on their cytotoxicity. Materials such as gold induce no reduction in cell viability over a wide concentration range (Sotiriou et al, 2014) and independent of the spatial configuration of the cells (Figure 1). Materials such as silver particles of relatively small crystal size (~8 nm) induce significant cytotoxicity in both configurations due to silver ion release from their surface (Sotiriou et al, 2014) (Figure 1). Finally, materials such as agglomerated silica particles induce higher cytotoxicity in upright than in inverted cell culture configuration, due to enhanced particle-cell contact in the former.

Cho, E.C., Zhang, Q. and Xia, Y.N. (2011) Nat. Nanotech. 6, 385-391.

Lee, S.H., Moroz, E., Castagner, B. and Leroux, J.C. (2014) J. Am. Chem. Soc. 136, 12868-12871.

Limbach, L.K., Li, Y.C., Grass, R.N., Brunner, T.J., Hintermann, M.A., Muller, M., Gunther, D. and Stark, W.J. (2005) Environ. Sci. Technol. 39, 9370-9376.

Madler, L., Kammler, H.K., Mueller, R. and Pratsinis, S.E. (2002) J. Aerosol Sci. 33, 369-389.

Pratsinis, S.E. (2010) AIChe Journal 56, 3028-3035.

Sotiriou, G.A., Etterlin, G.D., Spyrogianni, A., Krumeich, F., Leroux, J.-C. and Pratsinis, S.E. (2014) Chem. Commun. 50, 13559-13562.

Sotiriou, G.A. and Pratsinis, S.E. (2010) Environ. Sci. Technol. 44, 5649-5654.

Wengeler, R., Teleki, A., Vetter, M., Pratsinis, S.E. and Nirschl, H. (2006) Langmuir 22, 4928-4935.

Xia, T., Li, N. and Nel, A.E. (2009) Annu. Rev. Publ. Health 30, 137-150.

Copyright © American Institute of Chemical Engineers. All rights reserved.

488737 Association Rule Mining for Assessing the Relationships Among Biological Responses of Embryonic Zebrafish

Abstract:

The biological responses of embryonic zebrafish (including mortality) upon exposure to various nanomaterials were analyzed using association rule mining and clustering (hierarchical clustering and self-organizing maps (SOM)) analysis. The unique relationships (i.e., non-redundant association rules) among the various biological responses were identified from analysis of a nanomaterial biological interactions (NBI) knowledgebase. The biological responses were induced due to zebrafish exposure to seven different nanoparticles (carbon, cellulose, dendrimer, metal, metal oxide, polymeric, semiconductor) with average primary size range of 0.8 â€" 250 nm over an exposure concentration range of 0.016 â€" 55,000 mg/L. Pearsonâ€™s pairwise correlation analysis, along with hierarchical clustering and SOM analysis indicated that the non-redundant association rules were consistent with confirmatory tests. The analysis revealed that zebrafish olfactory regions are inter-associated and heart is inter-associated with touch response, circulation, delayed development, yolk sac edema, curved body axis, caudal fin, and pectoral fin. Spontaneous zebrafish movement was also found to correlate with yolk sac edema, axis and Somites. The results of the current study demonstrate that associate rules mining can be instrumental in suggesting means of optimizing the design of toxicity studies by streamlining the monitoring of multiparametric biological responses, improving understanding of the relationships among biological responses and for the development of nano-structure activity relationships.

Copyright © American Institute of Chemical Engineers. All rights reserved.

497786 Dynamic Nanoparticle Restructuring of Lipid Monolayers: Coating Amphiphilicity Trumps Charge

Abstract:

Langmuir monolayers formed by phospholipids deposited at air/water interfaces represent simple, but effective model biological membranes for studying interaction mechanisms between engineered nanoparticles and lipid membranes. This work describes the use of dynamic surface pressure measurements, coupled with optical and electron microscopy, to elucidate these mechanisms as a function of nanoparticle concentration and surface charge, and membrane composition, tension, and two-dimensional phase state. Saturated dipalmitoylphosphatidylcholine (DPPC, zwitterionic) and dipalmitoylphosphatidylglycerol (DPPG, anionic), and unsaturated dioleoylphosphatidylcholine (DOPC, zwitterionic) and dioleoylphosphatidylglycerol (DOPG, anionic) are used to form single or mixed monolayers with varying biophysical structure and phase behavior. Silver nanoparticles were examined with anionic, cationic, or PEGylated surface coatings because of their biological and commercial relevance. By varying the interlipid spacing, or lipid area per molecule, nanoparticle binding dynamics and the minimum insertion pressure were determined analogous to that for peptides or proteins. Unexpectedly, the PEGylated nanoparticles exhibited the greatest interaction with the lipid monolayers. Hydrophobic interactions, which were dominant for PEGylated nanoparticles and also present for charged nanoparticles, were responsible for monolayer insertion independent of monolayer charge. In contrast, electrostatic and charge-dipole interactions with PCs were primarily responsible for lipid condensation.

Copyright © American Institute of Chemical Engineers. All rights reserved.

493578 Liposomal Embedded Inorganic Nanoparticles and the Effects on Membrane Bending Elasticity and Structure Measured by Neutron Spin Echo

Abstract:

Nanoparticles (NPs) have unique optical, electronic and thermal properties and have found applications in biomedicine, gene therapy, targeted drug delivery, bioimaging, biosensors, catalysis and a plethora of other important applications. They are also an attractive option due to the straightforward synthesis, stability and precise control of their size and morphology during synthesis. The greatly enhanced use of these engineered nanoparticles has raised a question, whether there are potential health risks and side effects involved due to the unintentional interactions of biological entities with nanoparticles. The interactions between engineered nanoparticles and biological entities like biomembranes can also be intentional, as in the case of designing novel stimuli responsive drug delivery vehicles, imaging and diagnostics agents. Previous researches conducted in this regard have shown the toxic effects of NPs are dependent on the concentration, size, morphology, nature of surface functionalization and charges on the nanoparticles. Among the many toxic effects observed due to nanoparticle membrane interactions, formation of nanopores and unzipping of membrane bilayer are a few. It has also been shown that the membrane disruption due to nanoparticles depends on the phase state of the lipid bilayer. Toxicology studies have mostly been conducted on a symptomatic approach and there has been no clear consensus on the mechanism related to these effects. There is a dire need to study the interaction of these particles with cell membranes systematically and elucidate the interactions under various conditions of bilayer phase states and nanoparticle concentration, shape and surface functionalization.

The cell membrane comprises of a lipid bilayer and is the first biological entity encountered by the nanoparticle. The lipid bilayer comprising of phospholipids are ~5 nm thick. In this treatise, we have studied the interactions of gold nanoparticles (AuNPs) and phospholipid bilayers. We have not changed the surface functionality or morphology of the AuNPs, but only its mean diameter and concentration. We have focused on hydrophobic ligand stabilized gold nanoparticle having diameter less than the bilayer thickness (3 nm) and diameter more than the bilayer thickness (6 nm). A mixture of phospholipids (zwitterionic and anionic) was used to synthesize unilamellar vesicles in the 100 nm diameter size range. These vesicles are convenient models for studying cell membrane properties as most eukaryotic cells comprises of similar lipids. During synthesis of these vesicles in the presence of the hydrophobic nanoparticles, the 3 nm and 6 nm diameter AuNPs preferentially embed into the hydrophobic acyl region of the lipid bilayer. We hypothesize that the 3 nm AuNPs embed easily in the bilayer without compromising its integrity significantly, but the 6 nm AuNPs causes significant disruption of the bilayer due to a mismatch.

To answer our hypothesis, we have examined the bilayer thickness and membrane bending elasticity by utilizing Small Angle Neutron Scattering (SANS) and Neutron Spin Echo (NSE) spectrometry. Bending elasticity is a mechanical property that controls the thermal fluctuations of bilayer. We have attempted to explain the influence of AuNPs of different sizes and concentration on the bending elasticity of the model bilayer at a range of temperatures from the fluid to the gel phase. Most research focuses on measuring structure of bilayers by static methods only like SANS and x-ray scattering. NSE spectroscopy is the most suitable method for our studies because it is a dynamic method and is ideal for measuring thermal fluctuations in lipid bilayers because of its correlation times and length scales overlap with cell membrane fluctuations.

Copyright © American Institute of Chemical Engineers. All rights reserved.

497699 Natural Organic Matter and Bacterial Inoculum Concentration Affect Copper Toxicity to *Escherichia coli*

Abstract:

Copper-based engineered nanomaterials (Cu ENM), in the forms of copper hydroxide and copper oxide nanoparticles, are used in a variety of commercial products for agriculture. Copper has known antimicrobial effects, and usage of Cu ENMs can impact naturally occurring microbial communities. The primary source of Cu ENM toxicity is believed to be the copper ion. Here we consider its toxicity to *Escherichia coli* using growth inhibition experiments at various bacterial inoculation concentrations and with additions of humic acid (HA). We observe an inoculum effect for copper toxicity to bacteria where increases in initial bacterial concentration are associated with higher copper ion minimal inhibitory concentrations and minimal bactericidal concentrations. Furthermore, we find that HA at environmentally relevant concentrations can mitigate copper toxicity, requiring up to ten times higher concentrations of copper ion to have the same antimicrobial effect as when HA is absent. Bacterial growth rate constants generally decrease as ionic copper concentration increases. Copper ion binding isotherms are used to interpret results and estimate copper partitioning to *E. coli* and to HA. Transcriptional responses of three copper-responsive genes, *copA*, *cueO*, and *cusC*, were measured to identify changes in gene expression for different initial bacterial concentrations and the effect of HA when the same copper concentration is used for each system. Reactive oxygen species (ROS) were also measured to identify differences in the amount of ROS produced with different inoculum and HA concentrations.

Copyright © American Institute of Chemical Engineers. All rights reserved.

493894 Interactions of Engineered Sub-Micron Silica Particles with Cell Membrane Models

Abstract:

The greater prevalence of engineered nanoparticles in industrial and biomedical applications has led to increased exposure of mammalian cells to such particles. The plasma membrane, which delineates the extracellular surface, is the first cellular component that meets exogenous particles. A number of fundamental studies have used simplified membrane models to elucidate the interactions of nanomaterials with the cell membrane, but the role of nanoparticle physicochemical properties on these interactions remains obscure. In addition, it is unclear how the physical characteristics of the membrane model might affect the complex interactions of nanomaterials with lipid surfaces. The current study focuses on understanding the interactions of engineered silica nanoparticles with different surface properties, with two common membrane models, lipid monolayers and lipid vesicles, in an effort to elucidate how the functional groups on the surface of nanoparticles affect their interactions with different membrane models.

Lipid monolayers and vesicles comprised of an equimolar mixture of sphingomyelin, cholesterol, and 1,2-dioleoyl-sn-glycero-3-phosphocholine (SM/Chol/DOPC) were used as model membranes mimicking the outer leaflet and both layers of the cell membrane, respectively. Five silica particles with the same diameter (104 ± 5 nm), but coated with different surface-functional groups: hydroxyl, amine, and polyethylene glycol (PEG), with different molecular weights (2k, 5k and 20k) were used as the particle model, and were applied to the membrane models at three different concentrations (0.0001 g/L to 0.01 g/L). Studies with lipid monolayers were performed by spreading the lipids on a purified water subphase in a Langmuir trough and obtaining surface pressure isotherms by recording the surface tension using a Wilhelmy plate balance while compressing the monolayer. Nanoparticle effects on the integrity of lipid vesicles was studied by encapsulating the self-quenching fluorescent probe, 5(6)-carboxyfluorescein (CF), in vesicles and monitoring the release of CF from liposomes (vesicle leakage) upon exposure to nanoparticles. Fluorescence anisotropy, using the lipophilic probe, diphenyl-1,3,5-hexatriene (DPH), was employed to monitor the changes in lipid orderÂ in the cell membrane.

Surface pressure isotherms indicated that PEG-coated silica particles, regardless of their molecular weight, increase the packing of lipids, as evidenced by a rise in surface pressure, while amine and hydroxyl-coated particles did not significantly change the surface pressure isotherm. In contrast, amine and hydroxyl-coated particles induced a time-dependent leakage in lipid vesicles â€" demonstrating a loss of vesicle integrity. Unlike the monolayer studies, the effect of PEGylated particles on lipid vesicles was dependent on PEG molecular weight. While particles coated with PEG 20k induced significant vesicle leakage, particles coated with PEG 2k and 5k did not disrupt the vesicles demonstrating that PEG molecular weight regulates the disruptive effects of engineered particles. In addition, DPH anisotropy experiments revealed that vesicle leakage was not caused by particle entrapment in the lipid bilayer as no significant changes were observed in fluorescence anisotropy after exposure to the particles. In summary, these observations indicate that surface-engineered silica particles show differential effects on lipid monolayers compared to bilayers and their interaction is altered with changes in particle surface-functional groups. Future studies are focused on evaluating the effects of particles on the phase segregation of lipids by imaging giant unilamellar vesicles as the membrane model.

Copyright © American Institute of Chemical Engineers. All rights reserved.

503166 Nanocarbon-TiO_2 Composites for Photocatalytic Oxidation of Volatile Organic Compounds

Abstract:

Carbon nanotubes (CNTs), fullerenes, and reduced graphene oxide (RGO) have been coupled with TiO2 and the resulting photocatalytic activities have shown marked improvements. However, current design of nanocarbon-based TiO2 photocatalyst is still limited by the lack of in-depth understanding of their coupling mechanisms. Here, we investigate the photocatalytic activity of the CNT/TiO_2 and RGO/TiO_2 nanocomposite films in the degradation of acetaldehyde in a batch reactor under different conditions. In particular, the effect of initial pollutant concentration, catalyst film thickness, relative humidity, light intensity and wavelength, and type of substrate (and adhesion of film) on the photocatalytic degradation reaction of acetaldehyde will be presented. The expected synergistic effect and efficient separation of photogenerated electron-hole pairs in the synthesized nanocomposites have been verified. Our results show that the synergistic effect of nanocarbon is not evident in composites with physical interactions between TiO2 and nanocarbon. Interrelationships between properties of nanocarbon-TiO2 photocatalysts obtained from electron microscopic and spectroscopic characterization and photocatalytic activity have been established.

Copyright © American Institute of Chemical Engineers. All rights reserved.

500428 Nanobiosensing Chemicals That Target Nuclear Hormone Receptors: A Rapid, Versatile Cell-Free Protein Synthesis Approach

Abstract:

There is growing concern of the ubiquitous presence of endocrine disrupting compounds in our environment and their effects on human and animals (monitary impact estimated to be $340 billion USD in the US and $300 billion USD in the EU). This has catalyzed biosensor development to detect and study the interaction of these chemicals with nuclear hormone receptors (NHRs). Although progress has been made, available biosensors remain expensive and time consuming. Here we report a rapid, versatile cell-free protein synthesis platform for detecting chemicals that target NHRs. This biosensor is based on a multidomain-allosteric fusion protein with a core nuclear hormone ligand binding domain and a reporter domain. This biosensor has ability to produce a near real-time signal. We also report the robustness of cell free system in combination with a "just-add-sample" technique to detect endocrine disrupting compounds in environmental samples, including raw sewage. The cell free system has also been shown to be functional in human samples including blood and urine, which expands the possible portable applications of the biosensor.

Citation:

Salehi ASM, Shakali Tang MJ, Smith MT, Hunt JM, Law RA, Wood DW, Bundy BC. Cell-Free Protein Synthesis Approach to Biosensing hTRβ-Specific Endocrine Disruptors. Analytical Chemistry. 2017. 89(6):3395–3401

Copyright © American Institute of Chemical Engineers. All rights reserved.

496957 Detection of Biocides in Industrial Wastewater Using Pullulan-Encapsulated Microorganisms

Abstract:

Biocides, also frequently referred to as 'microbicides' or 'inhibitors', are used in industry to control or eliminate microbial growth in environments which are prone to microbiological activity. An evaporative cooling tower is an example of where biocides such as cetyl trimethylammonium bromide (CTAB) are deliberately used to control microbial growth to optimize cooling efficiency when cooling a building or a piece of machinery [1] [2]. Recent changes to water effluent quality standards are making the treatment of industrial wastewater increasingly difficult as municipalities are updating their sewer use by-laws to include restrictions on the amount of biocides. Therefore, there is a need for specialized tests to identify the presence of biocides in any potential incoming streams to municipal wastewater treatment plants (WWTPs) because the biocides are non-specific and thus will kill the beneficial microorganisms used in biological treatment processes (e.g. removal of ammonia). Existing methods for biocide detection rely on specialized techniques (e.g. High-Performance Liquid Chromatography, Liquid Chromatography – Mass Spectrometry) that are both time consuming and require the attention of specialized operators. This motivation has compelled our team to develop a robust, easy-to-use, biocide screening technology capable of accurately determining the presence of any biocide in wastewater. We envision that such a technology would be extremely useful for achieving 'point-of-use' reductions in the amount of biocides used at industrial facilities.

Using our expertise in the areas of WW separation processes and applied molecular biology, we have developed a bacterial cell-based kit that is capable of quantifying the relative toxicity levels of several WW samples in parallel in just minutes. The kit works by analyzing the survival rate of a library of microorganisms upon exposure to WW. For example, the survival rate of *DH5-α E. Coli* decreased from 85 ± 3% to 15 ± 3% as the concentration of CTAB increased from 60 to 120 μM. In order to obviate the need for cold-temperature storage, we have drawn upon our team's previous work and understanding in enzyme encapsulation using water soluble pullulan, and extended this to microbial cell preservation at room temperature. Pullulan is the polysaccharide used to make Listerine™ breath strips, and it has been reported as being extremely effective in protecting enzymes, RNA, DNA and small molecules against thermal and oxidative degradation [3]. In this new work, we performed an extensive optimization study that covered numerous experimental parameters (e.g. pullulan concentration, cell concentration, humidity) and storage conditions that resulted in a very robust and effective method of preserving microbial cells which are viable for long periods of time when maintained at room temperature conditions and higher (up to 40°C) - we have shown that *DH5-α E. Coli*cells can survive without refrigeration for long periods of time (>3 months), while maintaining survival rates as high as 50%. This works paves the way for creating a new area in sensing that uses whole cells as stable reagents that can be readily used in the field. In this work, we demonstrate this by showing that whole cells encapsulated in pullulan films can be used to detect the presence of the biocide CTAB.

References

[1] Baltimore Aircoil Company, "What is Evaporative Cooling?," 2010. [Online]. Available: http://www.baltimoreaircoil.com/english/what-is-evaporative-cooling. [Accessed 11 April 2017].

[2] S. Maruthamuthu, S. Ponmariappan, K. Indiral, A. Subramanian and N. Rengaswamy, "Interference Between Biocides and Inhibitors in Cooling Water Systems," Bulletin of Electrochemistry, vol. 16, no. 5, pp. 209-213, 2000.

[3] S. Jahanshahi-Anbuhi, B. Kannan, V. Leung, K. Pennings, M. Liu, C. Carrasquilla, D. White, Y. Li, H. R. Pelton, D. J. Brennan and D. C. Filipe, "Simple and ultrastable all-inclusive pullulan tablets for challenging bioassays," Chem. Sci, vol. 7, pp. 2342-2346, 2016.

Copyright © American Institute of Chemical Engineers. All rights reserved.

500426 Silica Nanoparticle Surface Characteristics Dictate *in vitro* Cytotoxic Behavior

Abstract:

Solid nanoparticles have proven useful in overcoming many of the barriers to drug delivery. Additional development has seen the attachment of a myriad of surface coatings such as polyethylene glycol (PEG). This commonly used PEG layer masks particles from innate host defence mechanisms within the body and increases circulation half-life of the therapeutic upon injection. Although this approach is widely accepted, the underlying mechanism is not well understood. This study examined the effect of hydrophilic surface coatings on solid silica nanoparticles with their interactions with Chinese Hamster Ovary (CHO) cells. Specifically, we examined the effect of differing Dalton weights of PEG on intracellular uptake and viability. Studies were conducted with amine and PEG 2k, 5k and 20k coated silica nanoparticles. Two nanoparticle sizes, 60 and 120 nm, were chosen to draw size effect comparisons.

In vitro distribution of the silica nanoparticles was found to directly correlate with their surface coating. Pegylated particles were found to remain interspersed among the CHO cells, whereas amine coated particles were found to produce punctate regions of fluorescence on and around the cells. Confocal microscopy confirmed uptake over simple surface association of the nanoparticles. Similar trends were correlated with flow cytometry. It was found that all PEG coatings decreased particle uptake by the CHO cells in comparison to those with an amine coating. The aminated particles had the greatest effect on cell growth. The 60 nm particles yielded greater half maximal inhibitory concentrations (IC_{50}) values over the 120 nm particles. Growth inhibition directly correlated with particle concentration regardless of surface character.

This study found that size and surface coatings have a dramatic influence on the CHO cell viability. Additional studies are needed to elucidate the interplay between both of these variable sets. This work has provided a basis of comparison that could be extended to various nanoparticle systems. Understanding the impact of widely used solid nanoparticle compositions, sizes, surface coatings and their combinations on mammalian systems is an important step towards developing robust drug delivery vehicles.

Copyright © American Institute of Chemical Engineers. All rights reserved.

501684 Development of Â Novel Nanocatalysts for Environmental Applications Using Helium Isolation Method

Abstract:

Nanoparticles (NPs) are revolutionizing many areas of science and technology, often delivering unprecedented improvements to properties of the conventional materials. However, despite important advances in NPs synthesis and applications, numerous challenges still remain. Development of alternative synthetic method capable of producing very uniform, extremely clean and very stable NPs is urgently needed. If successful, such method can potentially transform several areas of nanoscience, including environmental and energy related catalysis. Here we first present the experimental demonstration of catalytically active NPs synthesis achieved by the helium nanodroplet isolation method. This alternative method of NPs fabrication and deposition produces narrowly distributed, clean, and remarkably stable NPs. The fabrication is achieved inside ultra-low temperature, superfluid helium nanodroplets, which can be subsequently deposited onto any substrate. This technique is universal enough to be applied to nearly any element. We then prepared a series of model catalysts by depositing Au NPs on silicon wafers using helium nanodroplet isolation method. Taking the above-mentioned advantages of this method, we were able to study the support effect in catalysis by using these model catalysts.

Copyright © American Institute of Chemical Engineers. All rights reserved.

502111 Multifunctional Graphene Oxide As Remediation Agent

Abstract:

Graphene oxide has been extensively studied for a variety of applications as a result of its solution processing capabilities, rich surface chemistry, and facile reduction to reduced graphene oxide. The reduction process eliminates oxygen functional groups which in turn restores sp^2 domains and conductivity of the material. Little is known regarding the mechanisms involved in the reduction process and the final products that evolve from the oxygen functional groups tethered to the surface. Here, we demonstrate the multifunctional properties of graphene oxide as an aqueous remediation agent.

Copyright © American Institute of Chemical Engineers. All rights reserved.

493701 Procyanidin-Assisted Synthesis of Fe-Pd Nanoparticles for Superior Dechlorination of p-Chlorophenol

Abstract:

Nanoscale zero-valent iron has been widely applied in environmental remediation and toxic waste treatment. However, the dechlorination activity of Fe nanoparticles can be impeded by the iron hydroxides formed during the reaction. In this study, Fe-Pd nanoparticles (NPs) were synthesized and stabilized by procyanidin with superior dechlorination towards p-chlorophenol which can relieve the effect of the iron hydoxides. The prepared Fe-Pd NPs were characterized by scanning electron microscopy (SEM), transmission electron microscopy (TEM) and X-ray diffraction (XRD). Fourier transform infrared (FTIR) spectroscopy was used to investigated the interaction of Fe-Pd NPs and ferrous ions in the presence of procyanidin. The synthesized Fe-Pd NPs performed pseudo first order reaction rate constant. Increasing the amount of procyanidin in the synthesis of Fe-Pd NPs greatly enhanced the dechlorination activity, which may not only attribute to the homogeneous distribution of Fe-Pd NPs, but also attribute to the interaction of ferrous ions and procyanidin. The presence of procyanidin was found to avoid the aggregation of Fe-Pd NPs. Moreover, the formation of ferrioxyhydroxide on the surface of Fe-Pd NPs, which restricts the mass transfer and covers the active sites of Fe-Pd NPs, was also prevented by procyanidin. This strategy of the suppression of the formation of iron hydroxides on the surface of Fe-Pd NPs may be further applied to the recycling of Fe during the dechlorination.

Copyright © American Institute of Chemical Engineers. All rights reserved.

499015 Synthesis of Ni/γ-Al_2O_3 Catalysts for Methanation of CO_2 at Low Temperature Under Atmospheric Pressure

Abstract:

Wet incipient impregnation technique was used for synthesis of various Ni/γ-Al_2O_3 catalysts. The prepared catalysts were further used for methanation of carbon dioxide. It was noticed that the maximum conversion of carbon dioxide (87.5%) was achieved with 12%Ni/γ-Al_2O_3 at 300 ^{0}C, 6000 mL-g_{cat}^{-1}-h^{-1} and H_2/CO_2 molar ratio 4 under atmospheric pressure with 100% CH_4 selectivity. The effects of GHSV were also tested between the range of 6000 to 36000 mL-g_{cat}^{-1}-h^{-1}. The synthesized catalysts were also characterized by SEM, EDX, BET, XRD, TPR and TEM to check the surface morphology, elemental analysis, surface area and pore size distribution, phases of metals, reduction peaks and metal dispersion over alumina support, respectively.

Copyright © American Institute of Chemical Engineers. All rights reserved.

488573 Applications of Heterogeneous Fenton Catalysts in Pollutant Degradation, Sensing, and Cancer Treatments

Abstract:

Fenton process is a homogeneous reaction process from which strong oxidants, particularly hydroxyl radicals, are created through decomposition of hydrogen peroxide catalyzed by ferrous ions. These highly reactive species can oxidize and cleave large molecules into smaller ones. Consequently, Fenton processes have found extensive applications in pollutant degradation, particularly in wastewater treatment. This pollutant degradation process however suffers from simultaneous generation of unwanted side products, $Fe(OH)_3$, the so-called iron sludge. Heterogeneous Fenton process solved the above mentioned problem by using solid state iron-containing materials as the catalyst. The necessary redox reactions involved in the Fenton process is carried out with the lattice ferrous/ferric ion couple of the solid state catalyst, thus eliminating the iron sludge generation problem. In this presentation, several examples are offered to illustrate the application of heterogeneous Fenton catalysts in pollutant degradation. The essence of the heterogeneous Fenton process can be extended for use beyond pollutant degradation. We also demonstrate its applications in hydrogen peroxide sensing and cancer treatments, in which the heterogeneous Fenton reaction plays a key role.

Copyright © American Institute of Chemical Engineers. All rights reserved.

Application of Entrapped Nano Zero Valent Iron into Cellulose Acetate Membranes for Domestic Wastewater Treatment

Mohamed K. Mostafa[1], Ahmed S. Mahmoud[2], Rasha A. SaryEl-deen[2] and Robert W. Peters[3]

[1]Environmental Engineering Program, Zewail City of Science and Technology, Giza, Egypt

[2]Sanitary and Environmental Institute (SEI), Housing and Building National Research Center (HBRC), Giza, Egypt

[3]Department of Civil, Construction and Environmental Engineering, University of Alabama at Birmingham, Birmingham, AL

Abstract

Nowadays, nanotechnology and filtration using membranes have proven to be effective in wastewater treatment. Nanoparticles are highly reactive and effective redox and anti-microbial materials, while membrane filtration techniques reduce wastewater contaminants throughout the filtration process in an economical and fast manner. This study merges membrane technology and nanotechnology together through preparation of entrapped nano Zero Valent Iron (nZVI) into cellulose acetate (CA) membranes. The prepared nZVI was characterized by XRD and UV spectrophotometer, while the prepared membrane was characterized by measuring the water content, chemical compatibility, oil resistance, and membrane strength. The prepared membrane was tested for domestic wastewater treatment at different operating parameters pH, entrapped nZVI dose, and different flow rates. The efficiency of treatment was evaluated according to the reduction in turbidity, total nitrogen (TN), total phosphorus (TP), chemical oxygen demand (COD), biological oxygen demand (BOD), ammonia, nitrate, total solids (TS), total dissolved solids (TDS), total suspended solids (TSS), and iron concentrations. The associated reductions were recorded as 92, 70, 69, 80, 73, 37, 33, 36, 28, 77 and 64%, respectively for pH between 7 and 8 using nZVI dosage of 0.2 g/100 mL CA and a flow rate of 100 mL/min. This technique is more economical and efficient than the conventional technique.

Introduction and Background

Nanocatalysts technology, such as zero-valent metal have been widely used in wastewater treatment due to their higher reactivity and surface area which can degrade or adsorb different wastewater contaminants (El Saliby et al., 2008). The treatment processes using nanotechnologies depends mainly on the particle size. Several studies have been conducted to study the removal of wastewater contaminant using magnetic nanosorbents, such as nZVI. Contaminants that normally found in domestic wastewater includes inorganic cations and anions, as well as organic compounds. The organic compounds include chlorinated ethane, chlorinated hydrocarbons, poly and mono aromatic hydrocarbons and polychlorinated biphenyls which can increase the amount of COD in wastewater. Inorganic compounds include phosphate, nitrate, and sulfate (Harter, 2003, Philander, 2008, PEDDI, 2011, Wijesekara et al., 2014). Domestic wastewater may also contain some heavy metals, such as arsenic (As), lead (Pb), barium (Ba), cadmium (Cd), chromium (Cr), cobalt (Co), nickel (Ni), zinc (Zn), and copper (Cu). The bimetallic nZVI and nZVI are effective redox media that can be used as tertiary treatment for the removal of inorganic compounds from aqueous solutions. Entrapping nZVI in biopolymer has increased the validity of using nanoparticles in different applications, such as biomedical and biotechnological sciences because the biopolymers materials are non-toxic and non-immunogenic (Bezbaruah et al., 2009, Bezbaruah et al., 2011). Entrapping nZVI in cellulose acetate (CA) membrane has many advantages in water and wastewater treatment process, such as overcoming the separation problems of nZVI and increasing the pore size of CA membrane to reduce the pressure that used in the treatment process.

This work attempts to examine the removal efficiency of different contaminants found in domestic wastewater using entrapped nZVI into CA membrane. The relation between contaminants removal efficiency and the variation in entrapped nZVI dosage, pH, and flow rate were studied. This study also aims to prepare high quality and high strength membrane through preparing nZVI with particle size less than 50 nm.

Materials and Methods

Preparation of encapsulated nZVI into Cellulose Acetate membrane

Synthesis of nano scale Zero Valent Iron (nZVI); 2.1624 g $FeCl_3.6H_2O$ was dissolved in a 4/1 (v/v) ethanol/water mixture (96 mL ethanol + 24 mL deionized water) and stirred well. Also, 0.1 molar from $NaBH_4$ solution was prepared by adding 1.5128 g $NaBH_4$ in 400 ml of deionized water. Excess borohydride is needed to ensure better formation of nZVI. The $NaBH_4$ solution was poured in a burette and added at a rate of one drop per second into the $FeCl_3.6H_2O$ solution placed in a magnetic stirrer. After complete $NaBH_4$ solution the $FeCl_3.6H_2O$ solution turned to black precipitate. This precipitate was filtrated and washed using 99% ethyl alcohol. The black precipitate was dried at 75°C for one hour.

Synthesis of CA/nZVI membrane was carried out by phase inversion technique as follows; weight and dissolve 4g from cellulose acetate in 25 ml Dimethylformamide (DMF) solution at room temperature (22 ± 2°C). The CA was stirred for 60 minutes using magnetic stirrer at 300 rpm until complete dissolution was achieved. About 0.05 g of nZVI was added into dissolved CA solution with complete mixing, then stand 1 min before casting in 8 cm diameter Petri dish. Also gelation bath of 2L distilled water (non-solvent), containing 2wt% DMF and 0.2 wt% SLS was prepared and cooled to 17°. Then immersed the Petri dish in the gelatin bath to make hardness of solution.

Effect of operating parameters

The removal of different contaminants by CA/nZVI membrane was studied by batch technique. The removal efficiency for different contaminants was studied at pH: 3, 5, 7, 9, and 11, nZVI dosage: 0.2, 0.4, 0.6, 0.8, and 1 g/100 mL CA solution, and flow rates: 50, 100, 150, 200 and 250 mL/min.

Wastewater sample

Wastewater samples were collected from Orasscullia wastewater treatment plant (WWTP) which is located at New Cairo Governorate, Egypt. Samples were collected in 1-liter brown plastic bottles at 50 cm below the water surface. The storage solution for each test was identified according to ASTM 21 edition 2005. HORIBA instrument enabled measurement of pH, DO, TDS, and turbidity in the field. Analysis of other parameters took place in the Egyptian Housing Building Research Center (HBRC) laboratory, located in Cairo city. This laboratory received an ISO 17025 award for quality assurance. The samples were tested after 2 hours of collection according to ASTM standards. Table 1 includes analysis of the raw wastewater sample collected from the WWTP.

Table 1. Raw wastewater sample analysis

Parameter	Raw sample	Ministerial Resolution No. 44 of year 2000 (advanced)
DO, mg/L	2.2	$\geq$ 4 mg/L
pH	7.3	6-9
Turbidity, NTU	11.26	$\leq$ 20 NTU
COD, mg/L	202	$\leq$ 40 mg/L
BOD_5, mg/L	93	$\leq$ 20 mg/L
TN, mg/L	24	$\leq$ 15 mg/L
TP, mg/L	3.86	< 2 mg/L
Ammonia, mg/L	7.56	----
Nitrate, mg/L	0.9	----
TS, mg/L	623	----
TDS, mg/L	524	$\leq$ 2000 mg/L
TSS, mg/L	99	$\leq$ 20 mg/L
Fe, mg/L	0.28	----

Characterization of nZVI

XRD patterns were recorded at a radiation wavelength (Cu K alpha = 1.5418 A°). The X-Ray current and voltage values were 40 kV and 40 mA, respectively. The diffraction angle (2θ) ranged from 5° to 70° at a step size of 0.0167°(Xi et al., 2010). The UV-visible scanning spectrum was run for prepared nZVI after adding absolute alcohol (0.1 nZVI/20 mL ethanol) from wavelength 190 to 1000 with rate 50 nm/min (Abdel-Gawad et al., 2016). The prepared nZVI was scanned using scanning electron microscope (SEM) to discover the size and morphology of nZVI particles.

Characterization of the membrane

The water content (WC) of the prepared membranes was estimated by removing the membranes from water and weighing immediately after blotting the free surface water, and drying in oven for 1 h at 50°C. The percentage of water content of the membranes was calculated using Eq. 1.

$$WC = \frac{W_w - W_d}{W_w} \times 100 \qquad \text{Eq. 1}$$

where Wc is water content (%), W_d is dry weight of membrane (g), and W_w is wet weight of membrane (g)

The mechanical strength of the membrane was tested by measuring the tensile stress, elongation ratio, and young modulus values at break using Shmiduzue machine, model AG-X,100KN, Japan. The membrane was attached onto the 2 clamps of the machine and the draw speed was set at 10 mm/min. The correlation between stress (σ in MPa) and elongation (ε in %) was determined by Eq. 2.

$$\sigma = E\varepsilon \qquad \text{Eq. 2}$$

where E is Young's modulus (MPa) and ε equal $(L_o - L)/L$, L_o is original length, and L is the length after elongation

Acid resistance was studied using Hydrochloric acid (HCl) and Sulforic acid (H_2SO_4) using different concentrations. The membrane was immersed in acid solution for 1 hour. Solvent resistance was studied using ethyl alcohol, methyl alcohol, acetone, and benzene at different concentrations. The membrane was immersed in solvent solution for 1 hour. Oil resistance was studied using high and low carbon series oil. Fire resistance was also studied using direct fire for 7 min.

Results and Discussions

Characterization of nZVI

XRD analysis of prepared nZVI sample showed that the position of the diffraction peaks fit well with the body-centered structure of Fe mineral (JCPDS card No.87-0722). Two nZVI beaks was appeared at angles of 44.713° and 64.9° are attributed to (110) and (200) planes, respectively indicated that the prepared sample in crystalline structure and the powder sample free of oxides and impurities, as shown Fig. 1 (Mahmoud, 2017). By analysis the positions and the shape of the diffraction peaks, the inter-planar spacing (d) and the average particle size (D) of the prepared samples can be calculated using Bragg's law ($n\lambda = 2d \sin\theta$), where λ is the X-ray wavelength, n is an integer number and is equal 1,2,.... $n = 1$ in the above equation because for $n \geq 2$ the X-ray diffraction signal intensity will be much lower than for $n = 1$, and θ is the angle of deviation of the diffracted beam. From calculation the nZVI crystal size was between 23 and 59 and this results agree with SEM results.

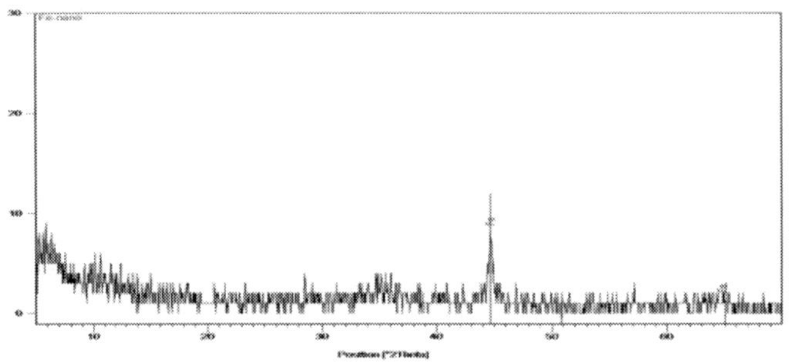

	Beta (deg)	Beta (Rad.)	2θ (deg)	θ (deg)	cos(θ)	Crystallite size D (nm)
Peak 1 (main)	0.384	0.0067	44.713°	22.35°	0.925	23.36999
Peak 2	0.166	0.002898	65.1°	32.55°	0.843	59.32

Figure 1. XRD of nZVI sample and related particle size

The Ultraviolet–visible spectroscopy (UV/Vis) results (Fig. 2a) indicated the formation of iron nanoparticles with particle size ranged from 10-50 nm due to the formation of main beak at high energy region of the scanning spectrum (Farahmandjou and Soflaee, 2015). Also, there is a small hump peak at 370 and 910 nm. A small hump at 370 nm confirmed the presence of Fe^{2+} ions with OH to form $Fe(OH)_2$ and other beak at 910 nm indicate the present of some oxides (Parveen et al., 2010).

Fig. 2b shows the SEM image which has been taken 1 day after preparation and it is clear that the prepared nZVI particles have formed large nanoclusters due to agglomeration. The nZVI particles also appeared as semi-spherical shape with an average size of 40 nm.

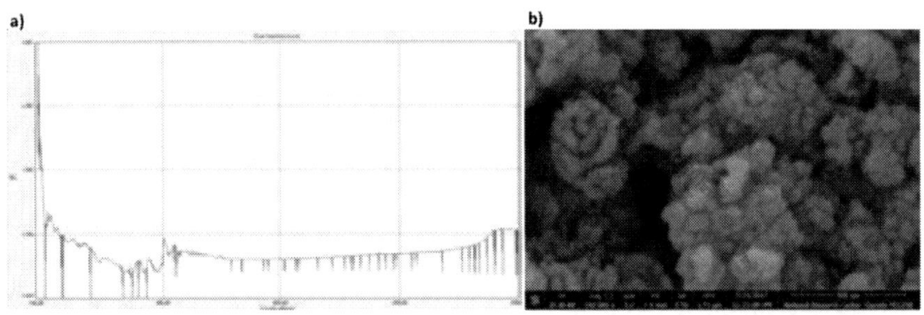

Figure 2. Characterization of nZVI: a) Uv-visble scanning spectrum and b) SEM analysis

Characterization of CA/nZVI membrane

Fig. 3a depicts the water content test of the prepared CA/nZVI membrane which is estimated as 60%. An increase in the water content is preferable in order to increase the service life of the CA/nZVI membrane, and thus decreasing the operating cost.

The mechanical strength was calculated for two types of membranes: (1) bare CA membrane; and (2) entrapped CA/nZVI membrane in order to study the effect of entrapping nZVI on the mechanical properties of the membrane. The stress, strain, and maximum force of the two membranes were measured at room temperature and their values are listed in Table 2. These results indicated that entrapping nZVI inside the membrane has improved the mechanical properties of the membrane, where the maximum force that the membrane can resist has increased from 10.5 to 10.9 N. Entrapping nZVI has also increased the tensile stress from 1.4 to 1.5 N/mm^2 and the strain from 86 to 116%.

Table 2. Mechanical strength results of bare CA membrane and entrapped CA/nZVI membrane

Membrane	Max. force, N	Tensile stress, N/mm^2	Strain, %
Cellulose acetate	10.5	1.4	86
CA/nZVI	10.9	1.5	116

The prepared CA/nZVI membrane showed high efficiency in resisting low and high series of oil (Fig. 3b). However, the membrane can adsorb methylene blue dies (Fig. 3c). The membrane can resist different organic solvent in low concentrations till 80% except acetone (Fig. 3d). The acetone solution can dissolve CA/nZVI membrane in very low concentration. High concentration of ethanol and methanol (exceed 90%) will close the active membrane pore size (Figs. 3e and 3f). The entrapped CA/nZVI membrane cannot resist sulfuric acid solution more than 50% (Fig. 3g). The effect of HCl on CA/nZVI membrane was studied using different concentrations leading to dissolve of entrapped nZVI at all tested concentrations (Fig. 3h). The prepared CA/nZVI membrane can also resist direct fire better than than bare CA membrane (Fig. 3i).

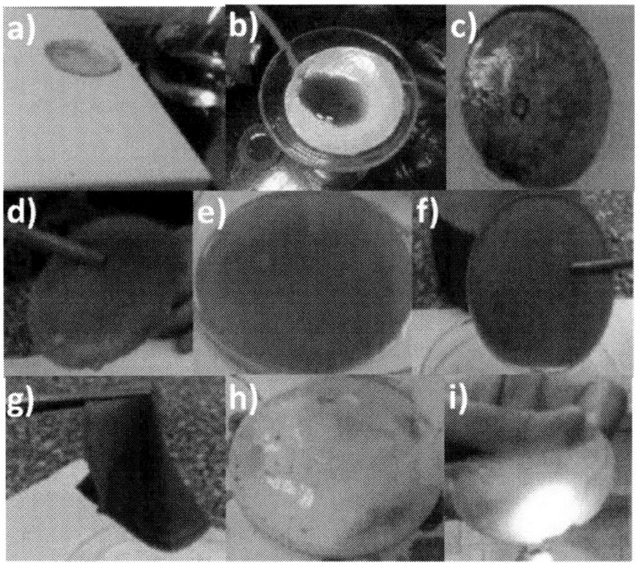

Figure 3. Chemical compatibility and fire resistance of CA/nZVI membrane

Effect of operating parameters

Effect of pH

The effect pH on different wastewater parameters removal efficiency was studied at entrapped nZVI dosage: 0.05 g, flow rate 100 mL/min, and temperature: 25 ± 3°C (Fig. 4a). The effect of pH was studied using different pH values 3, 5, 7, 9 and 11 on different wastewater parameters (Turbidity, TN, TP, COD, BOD, Ammonia, Nitrate, TS, TDS, TSS, and Fe). The removal efficiency was recorded as (95, 93, 92, 94, and 95%), respectively for turbidity, (60, 64, 69, 63 and 57%), respectively for TN, (64, 63, 68, 43, and 37%), respectively for TP, (76, 77, 79, 80, and 80%), respectively for COD, (73, 68, 73, 75, and 78%), respectively for BOD, (32, 34, 36, 17, and 19%), respectively for ammonia, (77, 55, 33, 11, and 11%), respectively for nitrate, (31, 33, 36, 37, and 39%), respectively for TS, (22, 25, 28, 29, and 29%), respectively for TDS, (77, 72, 76, 78, and 79%), respectively for TSS. This results illustrated that pH of 7 the most suitable media for the removal of these contaminants. The pH depends on the charge on the surface of the iron. The adsorption of negative and positive aqueous species depends also on the surface charge. When the surface charge is neutral it called point of zero charge (PZC). Unmodified ZVI is typically within pH 6 to 8 (Bezbaruah et al., 2009, Kosmulski, 2004). The concentration of the iron has increased in the acidic media due to dissolving the entrapped nZVI, however the removal efficiency increased in the alkaline media reaching 64, 39, and 42% for pH values 7, 9, and 11, respectively.

Effect of entrapped nZVI dosage

The effect of entrapped nZVI dosage on different wastewater parameters removal efficiency was studied at pH: 7 ± 0.2, temperature: 25 ± 3°C, and 100 mL of dissolved CA solution (Fig. 4b). The effect of nZVI dosage was studied at 0.2, 0.4, 0.6, 0.8, and 1 g and the removal efficiency was recorded as (93, 93, 94, 95, and 96%), respectively for turbidity, (70, 71, 72, 75, and 77%), respectively for TN, (69, 72, 77, 80, and 84%), respectively for TP, (80, 82, 85, 88, and 92%), respectively for COD, (73, 76, 81, 83, and 90%), respectively for BOD, (37, 43, 49, 51, and 57%), respectively for ammonia, (33, 41, 48, 54, and 63%), respectively for nitrate, (36, 36, 36, 36, and 37%), respectively for TS, (28, 28, 27, 27, and 27%), respectively for TDS, (77, 79, 82, 77, and 73%), respectively for TSS, and (64, 61, 46, 46, and 36%), respectively for iron. This results indicated that the minimum effective dosage for the removal of different wastewater contaminants was 0.2 g/100 mL. An increase in entrapped nZVI dosage increases the pore size of the membrane, and thus decreasing the removal efficiency of some contaminants (Fig. 4c).

Effect of flow rate

The effect of flow rate on different wastewater parameters removal efficiency was studied at nZVI dosage: 0.2 g/100 mL CA solution, temperature: 25 ± 3°C, and pH: 7 ± 0.2. (Fig. 4d). The effect of flow rate was studied at 50, 100, 150, 200, and 250 mL/min and the removal efficiency was recorded as (94, 93, 92, 92, and 85%), respectively for turbidity, (71, 70, 67, 64, and 62%), respectively for TN, (70, 69, 55, 49, and 46%), respectively for TP, (81, 80, 76, 72, and 65%), respectively for COD, (76, 73, 68, 59, and 51%), respectively for BOD, (41, 37, 31, 22, and 17%), respectively for ammonia, (47, 33, 18, 13, and 7%), respectively for nitrate, (38, 36, 31, 27, and 22%), respectively for TS, (30, 28, 25, 21, and 18%), respectively for TDS, (78, 77, 67, 58, and 49%), respectively for TSS, and (64, 64, 68, 71, and 75%), respectively for Fe. From the obtained result, the optimum flow rate was 100 mL/min. The removal efficiency decrease at high flow rates due to decreasing the contact time of through the membrane.

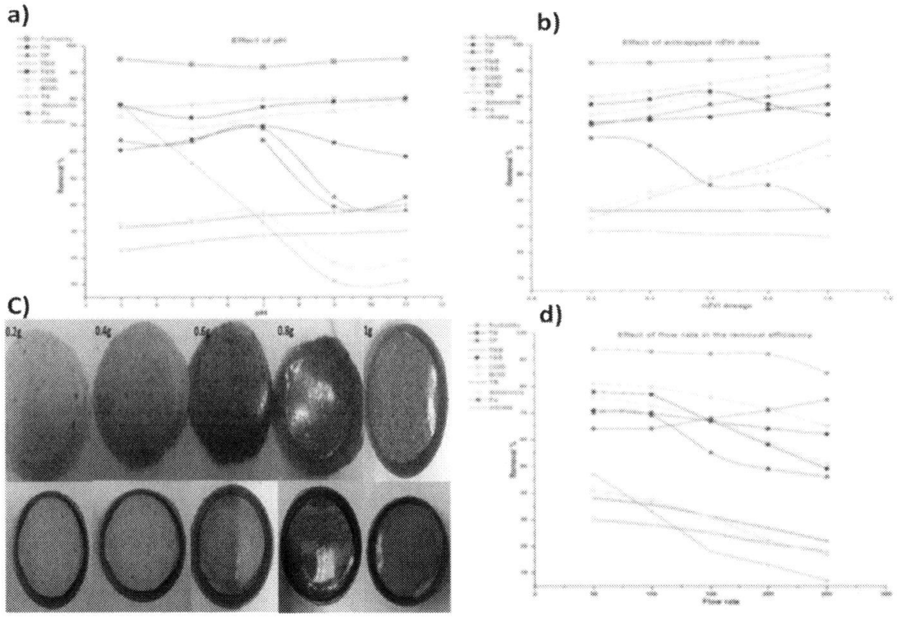

Figure 4. Effect of operating parameter on different wastewater parameters removal efficiency using CA/nZVI membrane: (a) pH, (b-c) nZVI dosage, and (d) flow rate

Table 3 lists the concentration of the studied parameters before and after treatment at the optimum condition of flow rate: 100 mL/min, pH: 7, and entrapped nZVI dosage: 0.2 g/100 mL. The concentration of all studied parameters has decreased significantly after using CA/nZVI membrane, and their concentration did not exceed the limits specified in Ministerial Resolution No. 44 of year 2000.

Table 3. Concentration of the studied parameters before and after treatment

Parameter	Raw sample	After treatment	Ministerial Resolution No. 44 of year 2000 (advanced)
DO, mg/L	2.2	2.5	$\geq$ 4 mg/L
pH	7.3	7.6	6-9
Turbidity, NTU	11.26	0.8	$\leq$ 20 NTU
COD, mg/L	202	41	$\leq$ 40 mg/L
BOD_5, mg/L	93	25	$\leq$ 20 mg/L
TN, mg/L	24	7.3	$\leq$ 15 mg/L
TP, mg/L	3.86	1.21	< 2 mg/L
Ammonia, mg/L	7.56	4.8	----
Nitrate, mg/L	0.9	0.6	----
TS, mg/L	623	398	----
TDS, mg/L	524	375	$\leq$ 2000 mg/L
TSS, mg/L	99	23	$\leq$ 20 mg/L
Fe, mg/L	0.28	0.10	----

Conclusion

This study has indicated that CA/nZVI membrane is highly efficient in removing different contaminants from domestic wastewater at 60% water content. Entrapping nZVI inside the membrane has improved the mechanical properties of the membrane. The prepared CA/nZVI membrane was efficient in resisting direct fire, low and high series of oil, as well as organic solvents except acetone. However, the CA/nZVI membrane was not efficient in resisting sulfuric acid solution more than 50%. The highest removal efficiency of all studied parameters was achieved at flow rate: 100 mL/min, pH: 7, and entrapped nZVI dosage: 0.2 g/100 mL. At the optimum conditions, the removal efficiency was 92% for turbidity, 70% for TN, 69% for TP, 80% for COD, 73% for BOD_5, 37% for ammonia, 33% for nitrate, 36% for TS, 28% for TDS, 77% for TSS, and 64% for iron.

References

ABDEL-GAWAD, S. A., BARAKA, A. M., EL-SHAFEI, M. M. & MAHMOUD, A. S. 2016. Effects of nano zero valent iron and entrapped nano zero valent iron in alginate polymer on poly aromatic hydrocarbons removal. *Journal of Environment & Biotechnology Research,* 5, 18-28.

BEZBARUAH, A. N., KRAJANGPAN, S., CHISHOLM, B. J., KHAN, E. & BERMUDEZ, J. J. E. 2009. Entrapment of iron nanoparticles in calcium alginate beads for groundwater remediation applications. *Journal of hazardous materials,* 166, 1339-1343.

BEZBARUAH, A. N., SHANBHOGUE, S. S., SIMSEK, S. & KHAN, E. 2011. Encapsulation of iron nanoparticles in alginate biopolymer for trichloroethylene remediation. *Journal of Nanoparticle Research,* 13, 6673-6681.

EL SALIBY, I., SHON, H., KANDASAMY, J. & VIGNESWARAN, S. 2008. Nanotechnology for Wastewater Treatment: in brief. *Encyclopedia of Life Support System (EOLSS).*

FARAHMANDJOU, M. & SOFLAEE, F. 2015. Synthesis and Characterization of α-Fe2O3 Nanoparticles by Simple Co-Precipitation Method. *Physical Chemistry Research,* 3, 191-196.

HARTER, T. 2003. *Groundwater quality and groundwater pollution*, UCANR Publications.

KOSMULSKI, M. 2004. pH-dependent surface charging and points of zero charge II. Update. *Journal of Colloid and Interface Science,* 275, 214-224.

MAHMOUD, A. S. 2017. Study the Removal of Most Hazardous Pollutants from Aqueous Solutions Using Nanoparticles. Cairo Univerisity.

PARVEEN, M. F., UMAPATHY, S., DHANALAKSHMI, V. & ANBARASAN, R. 2010. Synthesis and characterizations of nanosized iron (II) hydroxide and iron (II) hydroxide/poly (vinyl alcohol) nanocomposite. *Journal of applied polymer science,* 118, 1728-1737.

PEDDI, P. K. 2011. *PILOT-SCALE DEMONSTRATION OF hZVI PROCESS FOR TREATING.* Texas A&M University.

PHILANDER, S. G. 2008. *Encyclopedia of global warming and climate change*, Sage Publications.

WIJESEKARA, S., HARISCHANDRA, I., KUMARATHILAKA, S. & VITHANAGE, M. 2014. Fate and Transport of Selection Nutrients and Heavy Metals in Nanoscale Zero Valent Iron Amended Sand Columns.

XI, Y., MALLAVARAPU, M. & NAIDU, R. 2010. Reduction and adsorption of Pb 2+ in aqueous solution by nano-zero-valent iron—a SEM, TEM and XPS study. *Materials Research Bulletin,* 45, 1361-1367.

497602 The Role of Carbon Nanotubes in Photocatalytic Conversion of CO_2/H_2O and CO_2/CH_4 Systems to Syngas

Abstract:

The Role of Carbon Nanotubes in Photocatalytic Conversion of CO2/H2O and CO2/CH4 Systems to Syngas

Karishma Piler, Cristian Bahrim, Paul Bernazzani, Tracy J Benson.

The significance of photocatalysis in becoming one of the solutions in solving major air and water pollution problems has been increasing. In fact, NASA has been implementing photocatalytic technology in purifying air and water since 2011 and plans to use this technology in converting CO2 to fuels using solar light for In-Situ Resource Utilization (ISRU) Mars energy production. In addition, as part of Global Burden of Disease Project in Institute for Health Metrics and Evaluation, it was indicated that 5.5 million people die annually prematurely because of air pollution. Where in, air pollution caused primarily due to power plants, factories, and vehicle exhausts. Therefore, considering the significant role that photocatalysis plays currently and futuristically in air and water purification and alternative fuel production, it becomes increasingly necessary to optimize photocatalytic processes.

Our research work has led to the understanding and optimization of photocatalysis and photocatalytic processes through the synthesis of highly stable and highly efficient photocatalyst, chiefly for reducing CO2. The photocatalyst that we have produced contains single walled carbon nanotubes (CNT) doped onto titanium nanorods and have shown to convert mixtures of CO2/H2O and CO2/CH4 to synthesis gas (CO and H2) at low temperatures (100 – 150 °C) and pressures (1 – 3 atm) using visible light. Surfactants were employed to effectively disperse the carbon nanotubes. These dispersions allow for the de-bundling of CNT'S, providing large surface areas for other particles to come into contact. Therefore, the effects of three surfactants (SDS, SDBS, and CTAB) were studied to efficiently disperse carbon nanotubes with titanium dioxide nanoparticles. The synthesized catalysts were characterized using SEM, TEM, UV-VIS DRS, XRD, and FTIR. The reaction was carried out in an annular photochemical reactor using a 12 watt LED white light. The changes in CO2 concentration in the product gas were determined using GC/TCD.

In terms of optimizing the photocatalyst, the role of CNT in visible light driven photocatalysis was explored. There appears to be a synergistic effect which occurring between the TiO2/CNT that results in lower band gap energies (~2 eV). Implementation of the CNT'S have shown to have the following results for carbon dioxide conversion: 1) red shift of photocatalytic reaction, 2) Better adsorption of CO2 because of their nano-hollow structures including both SWCNT and MWCNT, and 3) longer lifetime of electron-hole separation because of its structure and large electron storage capacity.

Copyright © American Institute of Chemical Engineers. All rights reserved.

AUTHOR INDEX

AIChE
120 Wall Street, FL 23
New York, NY 10005-4020

ISBN 978-1-5108-5780-3

9 781510 857803

CURRAN ASSOCIATES INC.
proceedings.com